PLANTS

KT-430-537

Written by
Andrew Charman

Illustrated by
Martine Collings, Charles Raymond and John Shipperbottom

Edited by
Debbie Reid

Designed by
Charlotte Crace

Picture research by
Helen Taylor

CONTENTS

Plants for life

Plants are the most extraordinary things, even the ordinary ones. All kinds of amazing plants have developed over millions of years. There are now plants that can grow hundreds of metres high, live on other plants, catch insects and look like stones.

Every part of our lives is affected by plants. We could not live without them. They feed us, heal us, clothe us, transport us, protect us, and they give us the air that we breathe.

All the amazing features they have and incredible ways they behave are to help them survive and grow.

The bee orchid has flowers that attract bee pollinators.

Stone plants look just like stones and, in this way, they escape being eaten.

Rafflesia is the single biggest flower in the world. It is a parasite, taking its food from other plants.

Buttress roots may help to stop very tall rain forest trees from falling over.

A world of plants

Earth is a planet covered with plants. It is possible to put plants into a small number of very broad groups. To sort plants, you have to know how they are made, how they live and whether or not they produce seeds. Here are the main groups:

Algae

Seaweeds, which we we usually see on the beach, are kinds of algae. The green scum which forms on fish tanks, and the greenish-grey powder on tree bark, are also kinds of algae. Algae do not produce seeds.

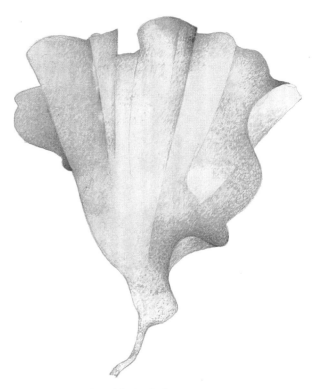

Sea lettuce is a kind of alga.

Fungi

The mushrooms and toadstools which we see in damp places in the autumn are fungi. They do not produce seeds. The part we see carries spores. These spores will be blown away in the air and eventually grow into new fungi. Yeast (used in baking) and moulds are also kinds of fungi.

The penny bun fungus – a type of fungi

Lichen

The orange or grey crust we often see on bricks, stones or logs is called lichen. There are thousands of different kinds of lichens and they are all made in a very unusual way. They are made up of two very different kinds of plant: a fungi and an alga. The alga makes the food from water, air and sunlight (see page 8), while the fungi gets food from whatever the lichen is growing on. The fungi also attaches the lichen to the rock or log. Again, lichen do not produce seeds.

Lichen growing on a stone

Ferns

Ferns are often found in damp, shady places, although some will grow in dry areas. Horsetails and club mosses are related to ferns. They all reproduce with spores.

Bracken fern

Mosses and liverworts

These small plants tend to grow in wet and humid places – in bogs and beside streams. Like ferns, they do not produce flowers or seeds, but spores.

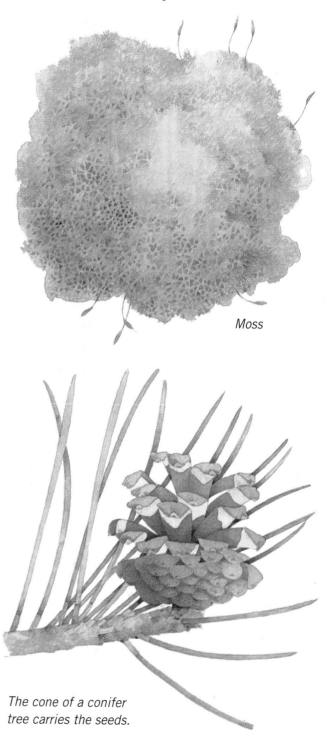

Moss

Seed plants

These plants reproduce with seeds. There are two groups of seed plants. One group, which includes conifer trees, have seeds which are not protected. The other group, the flowering plants, have seeds that are protected by a tough outer coating. Flowering plants are the most successful plants (see pages 6–7). There are nearly a quarter of a million different kinds.

Words to remember
Spores – tiny packages of cells that can grow into new plants.

The cone of a conifer tree carries the seeds.

The parts of a plant

No two kinds, or species, of plant are exactly alike. This is because they have developed different ways of surviving. Some may be tall, such as sunflowers; others, like daisies, can be short. Although they are different, their parts do roughly the same jobs.

The root

The root holds the plant in the ground and keeps it upright. The end of the root is continually growing and pushing its way between the particles of soil. Tiny root hairs absorb water and mineral salts from the soil. At the tip of the root is a hard cap which protects it from being worn away. Some plants have roots that can store food.

The stem

The stem holds up the plant so that its leaves catch as much of the sunlight as they can. It also connects the roots to the leaves.

At the top of the stem, this plant (opposite) has brightly coloured and scented flowers to attract insects. These insects carry out pollination (see page 12). This must happen if the plant is to produce seeds.

Leaves

The leaves take in light from the sun, carbon dioxide gas from the air, and water from the soil to produce food (see page 8). The upper surface of the leaf has a waterproof layer. This stops the leaf from losing too much water. On the under surface are pores (tiny holes) which control the gases going in and out, and how much water is lost.

Cells

All living things are made up of cells. Some plants consist of only one or a small number of cells. Most plants are made up of millions of cells. Cells can become specialized to do one particular job, such as carrying water or making food. To do this, the same kind of cells join together to form tissues.

Cell wall – holds everything together. May allow gases or water in and out.

Plastid – makes and stores food materials

Nucleus – controls what the cell does

Vacuole – fills with food or waste products, getting larger as the plant gets older

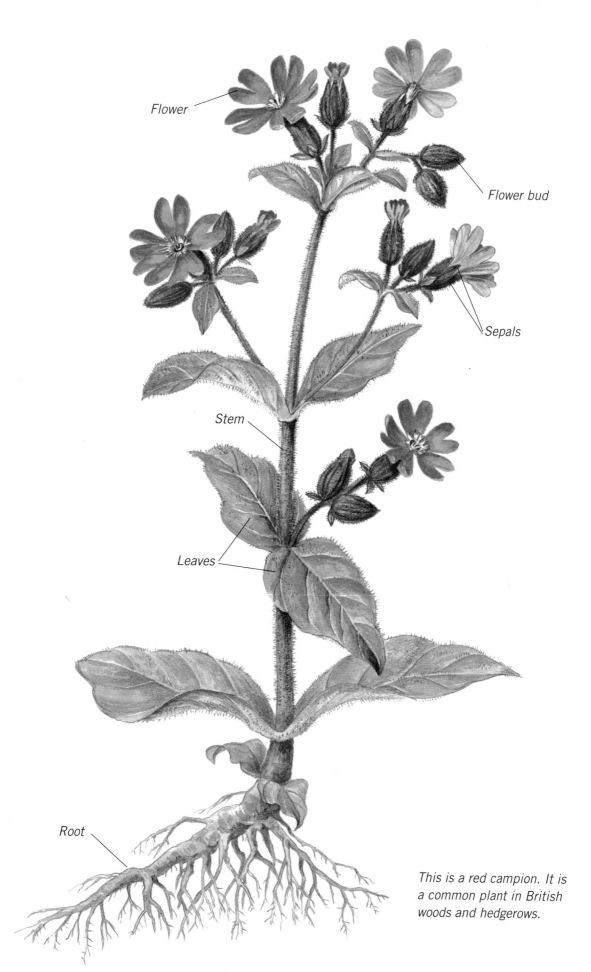

Flower

Flower bud

Sepals

Stem

Leaves

Root

This is a red campion. It is
a common plant in British
woods and hedgerows.

How plants feed

Plants don't look as if they do much, but all day long they are growing and reacting to their environment. The most important thing they do is make their own food. This is called photosynthesis.

Making food

Plants make their food from carbon dioxide and water, using light from the sun. Carbon dioxide from the air, and water from the soil pass into the leaves. When light falls on to the leaves, the gas and the water join together to make glucose. This is the plant's food which helps it to grow. Oxygen is given off by the plant at the same time (see page 46).

The need for healthy soil

To stay healthy, plants also need mineral salts from the soil such as calcium, magnesium, phosphorus and copper.

If the soil does not have enough of these minerals, the plants will not grow properly. We can give plants more minerals by adding fertilizers to the soil (see page 43).

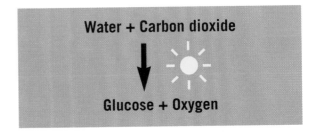

Water + Carbon dioxide

Glucose + Oxygen

Plants and water

Plants need water for photosynthesis. Water also helps them to stay firm – without it, they droop. Water dries, or evaporates, from pores (tiny holes) in the leaves. This water is replaced by more coming up from the roots.

Sunlight

Water

A plant makes its food from carbon dioxide and water when sunlight falls on to its leaves.

Death-trap plants

Some plants have an even more unusual way of getting extra food – they eat insects. The Venus' fly-trap and the pitcher plant are good examples.

These plants have special ways of catching unwary insects. The victim is trapped and digested, and the nutrients that make up its body are taken in by the plant.

The Venus' fly-trap catches insects in two jaw-like structures. The spines along the outside edges stop the victim from escaping before the jaws are tightly shut.

Taking from others

There are some plants that do not photosynthesise. They live on and get their food from other plants. These are called parasitic plants. Toothwort gets its food from the roots of the hazel tree. Only its flowers appear above the ground. A dodder plant twines itself around the stem of another plant to take food from it.

Common dodder smothering another plant

Growing and changing

Plants change as the year passes. They grow larger every day. Flowers bloom and die, and seeds are produced. Different seasons affect them too.

Growing

Unlike animals, plants keep growing throughout their lives. This growth not only makes them bigger, but it also enables them to move. Plants grow towards the light (see page 8).

Plants grow in height and their stems get thicker. At the growing tip of the shoot, the cells (see page 6) divide into two identical cells. These new cells then start to grow – they may get 1000 times bigger. This is what makes a shoot grow so quickly.

Growing wood

Trees and bushes grow woody stems. In the stem is a thin layer of tissue, called the cambium, which produces new cells. This makes the trunk get thicker.

Opening for the sun

Growing is not the only way plants react to outside conditions. Some, such as dandelions, open their flowers when the sun shines and close them when it does not.

Different life cycles

Turnip, radish, parsnip and carrot. These plants are biennials; the root vegetables are their food stores.

Turnip   *Radish*   *Parsnip*   *Carrot*

Plants live for different lengths of time. Annuals, such as peas and beans, take one year to complete their cycle. Biennials take two years. Carrots are a good example of this. The plant grows quickly in the first year and produces a store of food in the form of a big root – the carrot. We can pull up and eat this food store at the end of its first year. This store would be used to grow flowers and seeds the following year. Perennials live for three years or more. These include trees and shrubs.

Changing with the seasons

In the winter, it is too cold and dark for many plants to survive. Annuals die and their seeds wait for spring to come. Biennials lose their outer stems and leaves. Below ground, their bulbs or roots produce new shoots in the spring. Deciduous trees (see page 31) are perennials. They live for many years. In the autumn, they lose their leaves. They pass the winter as woody skeletons. In the spring, the temperature rises and there is more daylight. New leaves begin to grow because the tree can now take water from the ground which may have been frozen.

Spring

Summer

Autumn

Winter

An oak tree looks different during the four seasons of the year.

Plants and pollen

Just like animals, plants produce young. In flowering plants, male cells are transferred from one plant to the female cells in another. This second plant is now ready to produce seeds – it has been fertilized. These seeds will grow into new plants.

What is a flower?

There are many different kinds of flowers. They all have basically the same parts. The male parts and the female parts are usually found on the same flower. Some plants produce separate male and female flowers; other plants are themselves either male or female.

Insects and pollen

The colourful petals and sweet scent of flowers attract insects and other animals. These animals transfer the male cells, the pollen, to the female cells.

As well as colour and scent, some flowers also make sugary nectar. The insect visits the flower to feed on the nectar. As it feeds, it brushes against the anthers and pollen sticks to its body. It then visits another flower and some of the pollen is brushed off on to that flower's stigma. Fertilization can only happen if both plants are of the same species.

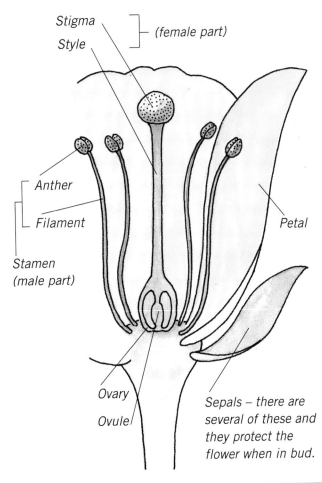

Stigma
Style
(female part)
Anther
Filament
Stamen
(male part)
Petal
Ovary
Ovule
Sepals – there are several of these and they protect the flower when in bud.

As the honey bee crawls about on the flower feeding on the nectar, its body is dusted with pollen.

Different kinds of flowers

For millions of years, plants have been developing different kinds of flowers to attract the insects' attention. Flowers may occur on their own or in groups. Plants have developed some special tricks to attract insects. Flowers that open at night to attract moths are pale in colour to reflect as much light as possible. Some flowers have lines of colour which guide the insects to the nectar.

From left to right: mullein, hogweed, wood meadowgrass and bindweed

Some plants attract just one kind of insect. The insect and the plant need each other to survive. Bats, small mammals and birds also carry out the pollination for some species of plants.

For some plants, an amazing trick is to look like the insect. The flowers of bee orchids, for instance, look like bees. A bee flies in and tries to mate with the flower. As a result, its body is dusted with pollen.

Words to remember
Fertilization – the stage at which male and female cells join together.

Using the wind

Most trees and all grasses are wind-pollinated. They produce huge quantities of pollen in the hope that some of it will reach the stigmas of other plants. As they do not need to attract insects, the flowers are small and dull. The stigmas are large for catching pollen.

Plants and animals

Plants do not live alone. They are part of communities made up of many different plants and animals. In any one place, the living things depend on each other for their survival.

Living together

Living things depend on each other and their environment for food, shelter and protection. A woodland may be home to thousands of different kinds of living things. No two live exactly the same way, and they all depend on each other. The birds need the trees and shrubs in which they build their nests. The flowers need the insects for pollination. The trees rely on animals to help spread their seeds. This means there is a 'web' of relationships between plants and animals. The study of plants and animals living together in the environment is called ecology.

Food chains

In a food chain, one living thing depends on another living thing for its food. For example, at the bottom of the food chain (shown opposite) are the plants. There are animals such as small fish which eat them. These animals are, in turn, eaten by other animals. As you go up the chain, the number of animals gets smaller. At the end of the chain are the the top predators; no one eats them.

What affects one part of the chain can affect other parts. For example, if you pollute the water, the fish will die and this will result in fewer herons.

Moving on to the land

Land can be stripped of all its plants. It can be damaged by pollution or covered by molten lava. If it is then left alone, the plants will eventually come back. The first to arrive will be algae, lichens and mosses (see pages 4 and 5). Other plants come in stages, each one becoming longer-lasting than the ones before it. Eventually, trees and shrubs will return. A woodland may take a hundred years to grow back.

Foxgloves are among the first plants to move on to a bare patch of ground.

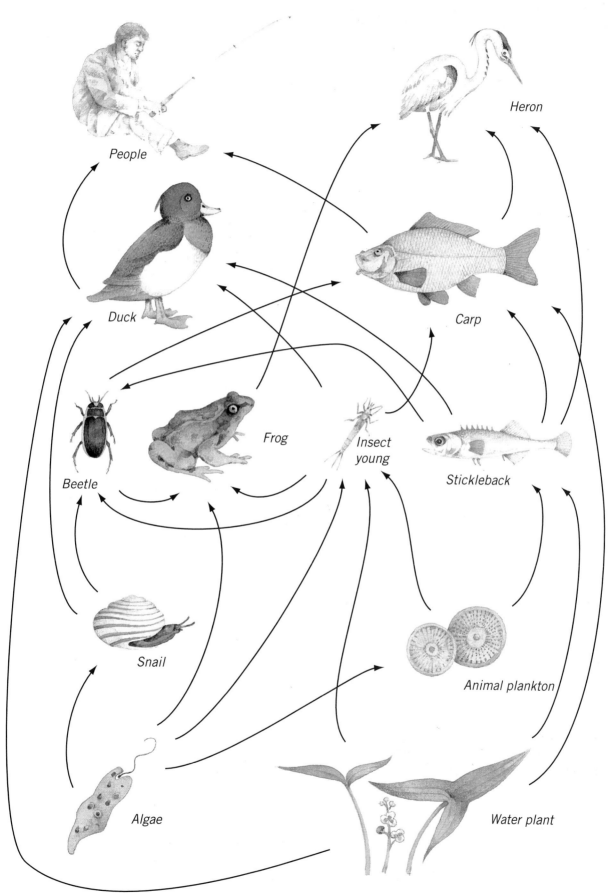

People

Heron

Duck

Carp

Beetle

Frog

Insect young

Stickleback

Snail

Animal plankton

Algae

Water plant

This is a typical food chain for a European lake.
It shows what eats what.

Making and spreading seeds

In flowering plants, seeds are formed when the plant has been fertilized (see page 13). New plants can grow from these seeds. Seeds can be very different shapes and sizes. They have many ways of moving away from the parent plant. This movement is called seed dispersal.

The fruit and the seed

After fertilization, the female parts of the flower (ovary) become the fruit. The ovule inside the ovary (see page 12) becomes the seed. The seed has parts which will become the root, stem and first leaves. The seed is protected by a tough outer coating, called the testa.

It is not always easy to tell which bit is the fruit and which the seed. With peas, for example, the peas are the seeds and the pea pod is the fruit. There may be just one seed, as with plums, several as with peas, or thousands as have been counted in poppy capsules.

Burdocks have seeds with spikes that stick to animals' fur.

Scattering seeds

Over millions of years, plants have developed many different ways of scattering their seeds. Plants use wind, water and animals to spread their seeds; some burst open and scatter their seeds around them. Whether those seeds will germinate or not depends on the conditions they find.

In dandelions, it is the ring of sepals around each flower that becomes the parachute. These fly away in the wind, each carrying a single seed.

Different fruits

There are many different kinds of fruits. They all have the same job – to be attractive and tasty to birds and animals. The fruit is eaten, but the seeds pass through the animal without being digested. The seeds may be lucky enough to land somewhere where they can grow.

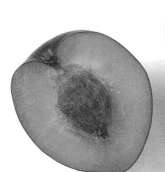

Plum

Plums are formed from a single ovary. The skin, flesh and hard 'nut' are the fruit. Inside the nut is the true seed.

Fig

Figs are formed from many ovaries of several flowers growing together as one unit.

Raspberries

These fruits are formed from about ten or twelve ovaries belonging to a single flower. Blackberries are the same.

Words to remember
Germination – early stages in the growth of a seed or spore.

Seeds starting life

With luck, some of the seeds produced by the plant will drop on to a patch of ground where they can grow, or germinate (see page 17). This patch will have nutrients, light and water. Even then, the seeds may not germinate immediately.

Sleeping seeds

The seeds may lie dormant (do nothing) for a while. In countries with warm summers and cold winters, like Britain, many plants release their seeds at the end of the summer. If the seeds germinated then, the seedlings would have to face a cold winter. So they wait, and grow in the spring when it is warmer and there are more hours of daylight.

Holding back the seeds

Many seeds contain a chemical which stops them from germinating immediately. This chemical is washed out of the seed by rain. This takes time – the time it takes for winter to pass, for example. Other seeds have a tough outer coating which takes time to wear away.

Seeds staying alive

The length of time seeds can live without germinating varies. Willow and poplar seeds can only survive for a few days. 'Weeds', however, can sometimes live for forty years! The seeds of the sacred lotus can live for 120 years, but the record is probably held by an Arctic lupin. Seeds that are ten thousand years old have come to life when given warmth and light.

Seed leaves

There are two main types of flowering plants. Woody trees, shrubs and many garden plants are called **dicots**. These have two seed leaves in the seed. Grasses, palms and orchids belong to another group called **monocots**. They have only one seed leaf (see page 19).

The eastern lotus is sacred to many people in India and Tropical Asia. It stands for purity, beauty, the sun and life.

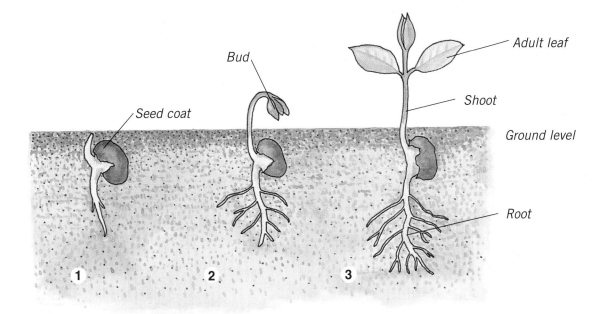

Growing

When conditions are right, the cells in the seed begin to divide and it grows.

1 The seed grows first at the root which breaks out of the seed coat. The root begins to take in water and mineral salts.

2 The shoot, or stem, often grows at first as a loop. This means that the delicate bud at the tip is not damaged as it pushes its way through the soil. Usually, the root grows downwards while the shoot seeks out the light.

3 The seed comes complete with seed leaves. In some plants, these stay below ground and provide food for the new growth. In others, they appear above ground and begin photosynthesis (see page 8). The seed leaves often do not look at all like the adult leaves that appear later. Once photosynthesis has begun, the seedling can feed itself and no longer depends on food from the seed.

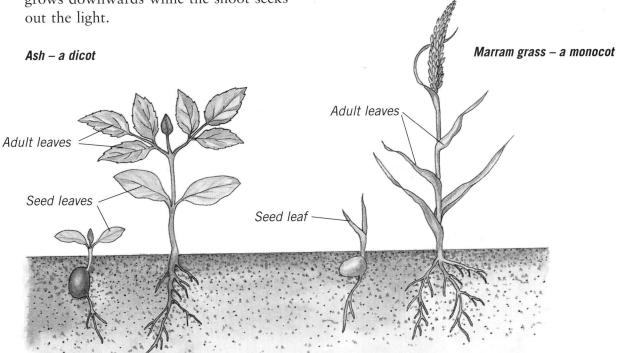

New life without seeds

Flowering plants have another way of reproducing themselves. This can take place without flowers and seeds. What happens is that part of the plant becomes a new plant. This is called vegetative reproduction.

Identical plants

Most flowering plants can use vegetative reproduction as well as flowers and seeds. Vegetative reproduction avoids pollination and germination, both of which are risky and may not happen. The problem with this method, is that the new plants are exactly the same as the parent. If conditions change and the parent dies, so do the new plants. If the new plants have been formed from a seed and are slightly different from the parent, they have a better chance of surviving.

Stem growth

Often the new growth comes from buds on the stem. Potatoes are known as stem tubers. They are a store of food and each potato can grow into a completely new plant. Some bulbs, such as crocus bulbs, reproduce themselves every year. If you break off the new bulbs and plant them, they will grow into new plants.

The house leek produces a ring of 'offsets' on stem runners. Each of the offsets can become a new plant.

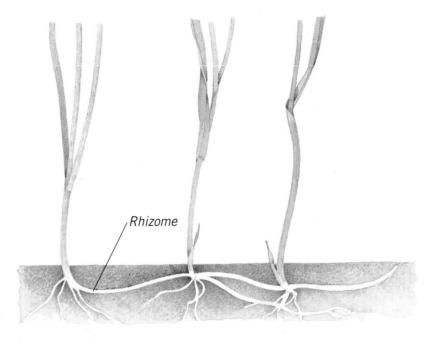

After they have produced their fruit, strawberry plants grow leafy stems called runners. These run along the ground.

Runner

Branching out

Some new plants grow from buds on stems which stretch out along the surface of the soil or just below it. They are called runners. Strawberries and buttercups grow like this.

Rhizomes are stems that grow under the ground. Bindweed and couch grass grow like this. New plants grow up at certain points along the stem. Once the new plant has roots and leaves of its own it can grow without the parent plant.

Rhizome

Vegetative reproduction makes couch grass a stubborn weed.

Plants worldwide

Plants live in almost every part of our planet. It is only in the hottest and coldest places that they cannot survive.

The vegetation zones

Which plants grow where in the world depends on the climate (weather patterns). This affects the temperature, the amount of light and water, and what the soil is like. Across the world, there are certain vegetation zones that we can easily identify. Particular plants have adapted to survive the conditions in each zone.

Some live in only one zone, others are more widespread.

No two species of plant live in exactly the same way. Some plants like shade, some like wet soil – they are all different. Being different means that they do not all want the same things. This increases their chances of survival, and they do not need to compete with each other.

In each vegetation zone, there are certain plants that are dominant. They take most of the light, water and nutrients. Often these are the trees.

> **Words to remember**
> Temperate – places which have warm summers and cold or cool winters.

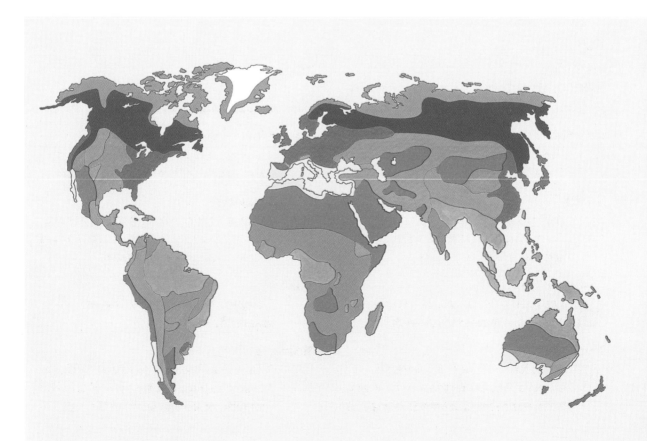

This map of the world shows where the different vegetation zones can be found. Use the key on page 23 to help you.

The open grasslands

There are two different kinds of grassland in the world – the savannah and the steppe. Savannah grasslands are found in dry, tropical areas in Africa, South America and Australia. In Africa, the grassland is lush in the rainy season and dry at other times. It is dotted with acacia trees which give shade to elephants and lions. The grass is grazed by huge herds of plant-eating animals such as wildebeest and zebra.

Steppe grassland is found in temperate areas of North America, Russia, South America and Australia. In America, it was once the home of wandering herds of bison. Large areas of these grasslands are now used for growing crops.

A herd of wildebeest on the African savannah.

Key

Hot deserts
Only plants that can survive great heat and very little water live in these zones.

Cold deserts
 Here, it is too cold for plants to survive. The water is always frozen.

Tundra
It is cold, but there is a little water. There are short shrubs, mosses, grasses and some flowering plants.

Coniferous forests
The trees here are mostly evergreen with needle-like leaves (see page 30).

Temperate forests
These forests have great seasonal changes. Deciduous trees such as oak, birch and beech live here (see page 31).

Tropical rain forests
These are the lushest and most varied vegetation zones in the world.

Temperate grasslands
 These are flat, sandy plains. Rain drains away quickly so that trees cannot grow.

Tropical grasslands
 These are home to great herds of grazing animals. In places, the grass is mixed with thorny scrub or open woodland.

Scrub
 Evergreen trees or shrubs, or open grassland.

Mountains
 Mountains often have forests at their base, tundra-like plants higher up, and usually just rock, or ice and snow, at the top.

Plants in water

The sea is the largest habitat on Earth. It is home to millions of creatures. Just as on land, plants photosynthesize to make food.

The plants that use most of this light are tiny blue-green algae. They are then eaten by plant-eaters, such as krill, who are eaten by meat-eaters including whales and sharks. So plants are the source of all food in the sea.

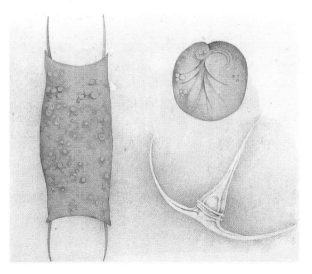

Tiny plants called phytoplankton drift in the surface layers of the sea where it is warm and the light can reach them.

Plants on the beach

The seaweeds that we see on the beach at low tide are also algae. They live on the edge of the sea where they are covered by water at high tide and exposed at low tide. They have tough outer layers to prevent them being dried out by the sun. Seaweeds have holdfasts (instead of roots) which fix them to the rocks. They are flexible plants and can bend with the water. They are not easily damaged.

A diver swimming through a kelp forest. Kelp is one of the largest seaweeds. It can grow to a length of 60 metres.

Further up the beach

Some land plants are able to live near the edge of the sea. They can grow where there is a lot of salt. Marram grass grows long, spreading roots which bind the sand. This stops the sand being washed or blown away, and eventually makes land for other plants to grow on.

Freshwater plants

Plants cannot take root in fast-moving rivers, but, as the current slows down, they can take a hold. Some river plants, such as fanwort, have fine leaves which are not damaged by the current. Many water plants have their roots in the soil at the bottom but show their flowers above the water. They can then be pollinated by insects. Water lilies use the surface of the water to support their huge leaves.

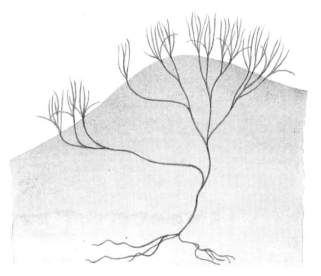

The long roots of marram grass seek out water. They bind the loose sand so that it builds up into dunes.

Words to remember
Habitat – the conditions in which a living thing lives.

The leaves of the Amazonian water lilies can reach a diameter of more than 2 metres. They are strengthened by strong ribs on the underside.

Life in the cold

In the world's cold places, there are sharp, icy winds. There is not very much rain and what does fall may come down as snow. Any water in the ground is likely to be frozen. Plants have found ways of surviving all this.

Mountain survivors

The higher you travel up a mountain, the colder the air becomes. Plants that live on high mountains are called alpine plants. They are usually small and compact, growing in thick clusters and flattened mats. This stops them from being damaged by cold winds. Their leaves are small which means they do not lose too much precious water.

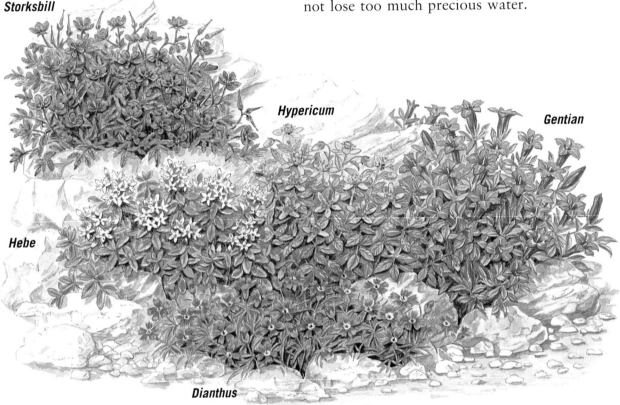

Storksbill

Hypericum

Gentian

Hebe

Dianthus

A collection of mountain, or alpine, plants.

The silversword grows on the mountains of Hawaii. Here, near the tropics, the sun is hotter and plants are exposed to dangerous ultraviolet rays. The silversword is protected from these rays by fine white hairs that grow on its leaves.

The dwarf hebe grows on mountains in New Zealand. It forms thick, low-lying cushions. These trap heat and prevent damage by the wind, especially water loss. The leaves are tough enough to withstand sharp frosts.

In the mountains of Central Africa and South America live plants of great size. Lobelias and groundsels grow several metres high. It is not known exactly why. In other parts of the world, these plants are quite small.

Life on the tundra

On the edge of the icy Arctic, around the North Pole, is an area known as the tundra. Roughly the same kind of plants grow here as in the alpine zone of mountains. The tundra land is covered with snow during the long, dark winters. When summer comes, a thin top layer of soil thaws out. Lichens, algae and mosses are able to grow. Also, there are a hundred species of flowering plants such as cranberries, stitchwort and dwarf birch. Trees cannot grow – it is too cold and there is not enough water. The Antarctic, around the South Pole, has few plants because it is too cold. Lichens grow on the rocks, and mosses grow where there is soil.

In some mountain areas, the summers are short and the winters long. The plants bloom quickly and at the same time.

Life in the heat

There are some large areas on Earth where very few plants can survive. These are the deserts. Here, there is very little water for the plants to use. The frozen expanses of the poles are deserts (see page 27). So, too, are the Earth's hot spots.

Desert survivors

Some plants have found ways of surviving in the hot deserts. Some, such as lichens, can simply live for a long time with very little water in their tissues. The creosote bush just shuts down in a drought. Its mature leaves drop off and its buds turn brown. They stop photosynthesizing. When the rain comes, they turn green again and life continues.

The creosote plant stops growing during long periods of drought. It comes to life again when the rain falls.

Searching for water

Keeping water in is useful; so is being able to absorb every last drop of what does fall. Many desert plants have shallow roots that spread out over a very large area. Others send down deep roots to reach water.

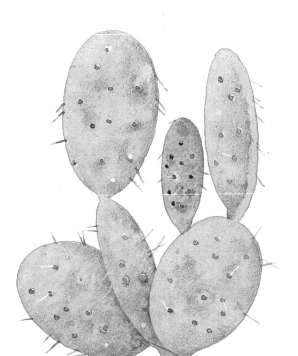

Cactus' roots do not go very deep. They spread out over a wide area to collect as much rain as possible. This is a prickly pear cactus.

Keeping the water in

Plants lose water through the pores (tiny holes) in their leaves. Water also evaporates from the surface. Most desert plants have their pores on the underside of the leaves to reduce this loss. Other plants have very small leaves or none at all and use the stem to carry out photosynthesis (see page 8).

Storing water

Cacti are plants that can store water (succulents). They use their thick stems as containers. They have no leaves but lots of spines instead. This means that the plant does not have broad, flat areas that can be dried up in the sun.

The cactus' spines may protect the plant from the extremes of hot and cold and from attack by animals.

Water in leaves

Some succulents, such as the necklace vine, use their leaves for storing water. Others use their roots. Leaf succulents live in slightly damper areas than those in which cacti can survive. Their leaves have a thick, waxy surface which protects them from the harsh sun and helps them to stay cool.

Blooming quickly

Some desert plants survive the long periods between rain storms not as plants, but as seeds. They may lay dormant, doing nothing for years. When the rain comes they bloom, are pollinated and set seed very quickly before the soil dries up again. There may be as many as 25 000 seeds per square metre just waiting for rain.

During periods of drought, a leaf succulent's leaves become shrivelled and wrinkly. They swell up with stored water when the rains come. This plant grows in Southern Africa.

Forests

Forests are communities of plants – the dominant plants are trees. Where light falls between the trees, smaller plants can survive. Some can even live on the trees themselves.

What is a tree?

A tree is a plant with a single woody stem. It is a material called lignin which makes the stem hard and strong. This enables trees to grow taller than other plants in the contest for light. The most common trees are the broadleaves, such as oak and ash; and the conifers, which include firs and spruces.

Coniferous forests

The northern parts of Europe, North America and Asia have coniferous forests. Spruces, firs, pines and cedars are the most common trees. They are evergreen – they do not lose all their leaves at once in autumn. These trees can survive in places where the winters are long, cold and dark. Their leaves are tough, leathery and needle-like and do not dry out in the cold winds.

In sunny gaps, shrubby plants such as bilberries and junipers survive. Grasses and plants such as wild strawberry can also be found. Mosses and lichens grow on the trees, and fungi live on the forest floor.

Coniferous trees cast a dense shade. Only when a tree dies and leaves a gap can other plants thrive on the forest floor.

Temperate forests

In more southerly parts of Europe, North America and Asia, there are temperate forests. Here, the summers are warm and the winters are cold. Broadleaved trees, such as oak, ash, beech, maple and birch grow in these forests. During the summer, the trees have leaves which cast a dense shade. In the autumn, they lose their leaves and lie dormant over the winter. These are called deciduous trees.

In the early spring, the trees do not have any leaves and there is less shadow. Many smaller plants grow in the light that falls on them. These include wood anemones, wild strawberries, primroses and bluebells.

Key

① wild strawberries
② toadstools on moss
③ bilberries
④ chickweed
⑤ cowberries
⑥ Scots pines
⑦ juniper
⑧ Lady's tresses
⑨ fungus
⑩ lichen

This is a rain forest in Venezuela, South America. Rain forest trees are evergreen, losing a few leaves throughout the year rather than all at once.

The crowded rain forests

The tropical rain forests are packed with plant life. Conditions are right for growth all year round: heavy rainfall, warm temperatures and long hours of daylight. The trees cast a dense shade with their dark green, glossy leaves. Taller trees break free of the shade, and below them thrive a wide variety of shrubs and smaller plants. Some plants grow on the trees themselves. These include lichens, mosses, liverworts, ferns and orchids. The trees are entwined with woody-stemmed climbing plants called lianas. These have their roots in the soil and develop a huge spread of leaves.

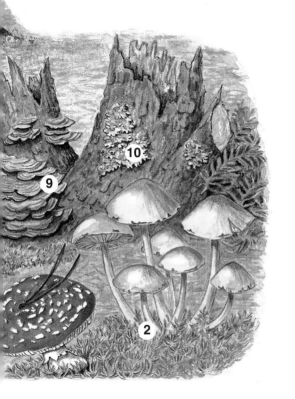

Plant defences

Most plants are rooted to the spot. Unlike animals, they cannot run away from their enemies. Fortunately, plants have different ways of defending themselves, such as fine hairs, stings or stabbing thorns.

Once stung, twice shy

Stinging nettles can be found in hedgerows and rough patches of ground. The stings of the stinging nettle are like tiny syringes. They are strengthened by a glass-like material called silica. When an animal brushes against the plant, the stings pierce the skin. They inject a chemical which causes a painful rash.

Hairy plants

Insects crawl on to plants and suck nutrients out of the stems; some chew great holes in the leaves. The leaves of some plants are covered with fine hairs. Small insects find it difficult to walk across the surface, and so they give up and go elsewhere.

A stinging nettle sting enlarged many times.

Stinging nettles grow where the soil is rich in nutrients.

Ants to the rescue

Some acacia trees have ants to protect them. The ants live on the tree, feeding on the sweet-tasting pith (flesh) inside the thorns and on the leaves. In return for this food and shelter, the ants attack any animal that attempts to eat the leaves.

Dangerous chemicals

Many plants have substances in their leaves and stems which make them unpleasant or dangerous to eat. These dangerous chemicals have been used to make poisons. The South American Indians hunt with a poison called curare. This poison is now used in a safe form by surgeons to relax patients' muscles.

Indians from Ecuador make the curare poison from the bark of a vine.

Spine attack

Spines, thorns and prickles are common forms of self-defence for plants. Briars and wild roses have sharp thorns on their stems which can be very painful to an animal or person.

Thistles protect themselves with prickles, which can be very painful if you try to pull one up without wearing thick gloves!

A watery trap

Troublesome snails and insects can drown trying to get to the leaves of the plant called teasel. Where the pairs of leaves join the stem of the plant there is a pool of water. Insects trying to get to the leaves run the risk of a watery grave.

Food for all

Plants have been used for food and flavourings since the first people felt pangs of hunger. Over 10 000 different kinds have been eaten at one time or another, including fungi and seaweed.

Staple foods

Today, only a small number of crops provide the world with most of its food. These include rice, wheat, maize, barley and sugar cane. They are the world's most important crops and are known as the staples. Which crop is grown where depends on the climate. Rice is grown in tropical countries. Wheat is more common in cooler countries, such as Britain.

Apple

Chard

Cassava

Olive

Orange

Star apple

Kidney beans

Chickpeas

Swede

The many kinds of plant food we eat come from different parts of the plants. There are stems, leaves, seeds, fruits, even flowers.

Other important crops are the swollen roots and stems of many plants. These include potatoes, cassava, carrots, swedes, onions, turnips and many others.

When we eat pulses, we are enjoying the edible seeds of the pea family. Soya beans, lentils and chickpeas are all kinds of pulses. These are good sources of protein.

We also harvest the fruits of many plants including grapes, bananas, pineapples, tomatoes, oranges and apples. We eat the green leaves of cabbages and spinach, and the unopened flower heads of broccoli.

Oils from plants are also used in cooking. Oil crops come from olives, the oil palm, sunflowers, sesame seeds and coconuts.

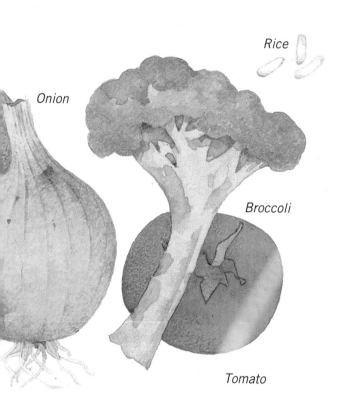

Rice

Onion

Broccoli

Tomato

Old crops and new crops

For thousands of years, farmers have been improving their crops by only planting the seeds from their best plants. We now have crops that produce more food than ever before. It is hoped that these crops can help with food shortages across the world. Unfortunately these 'megacrops' can easily be attacked by disease and they need expensive fertilizers, machines and lots of water to grow.

Another future for plants

Some people believe that there is another way forward. This is to grow plants that can survive in harsh conditions. There are many such plants flourishing in the wild. Some like dry weather, others succeed in soil which is too salty for other crops. For example, the buffalo gourd grows in the very dry parts of Mexico and parts of America. It can be 3–4 metres long, and its fruits produce seeds with lots of protein and oil.

Jackfruit plants grow well in salty soils. The fruits can be boiled, roasted or eaten fresh. This one is from Southern India.

The luxuries of life

There are a large number of plants that give us our luxuries. They make up the drinks we enjoy; the herbs and spices that add variety to our diets; the perfumes, cosmetics, shampoos and bath oils we use to make ourselves clean or just more attractive.

Plants we drink

Probably the most important plants we use as drinks are tea and coffee. Both have been enjoyed for thousands of years. Tea is made from the fresh young leaves of a plant which is grown in hot countries, particularly India. Coffee is made from the roasted beans (seeds) of a tropical evergreen shrub. Both tea and coffee are grown on large plantations.

Chocolate is made from the beans of the cocoa plant. It was first used by the Aztecs of Central America.

All kinds of alcoholic drinks are made from plants. Barley, rice, grapes, molasses, and even potatoes, are commonly used. They are flavoured with hops, juniper berries, aniseed and many other plants.

Flavouring food

Herbs have leaves with pleasant flavours – they add interest to food. In medieval times, before refrigeration, they were used to disguise the flavour of rotting food. They came originally from Europe. Among them are thyme, rosemary, basil, marjoram and oregano – they are all members of the mint family.

Spices come from tropical woody plants. Many are the seeds of trees. Cinnamon comes from the bark of a tree; vanilla comes from the seeds of a climbing orchid. Nutmeg and mace come from the same tree – nutmeg from the seeds, mace from the fleshy part of the fruit.

Rosemary

Thyme

Marjoram

Basil

Oregano

Mace

Vanilla

Nutmeg

Cinnamon

Plants for make-up

All over the world people use make-up to enhance their looks. Modern make-ups are rarely made entirely from plants, but some people do still use natural make-up. The Amazonian Indians make one from the waxy coating of the seeds of a tree. The seeds are made into a red paste – deep scarlet for the men, a paler orange for the women. The indians like to wear bright make-up because they hate drabness – it makes them think of death.

Plants in the bathroom

We use many plants in our bathrooms. Soaps, shampoos, moisturising creams and make-up all contain plant oils. These plants include soya beans, sunflowers, coconut palms and olives. Toothpastes are often flavoured with mint, fennel or cloves.

Sweet smells

One of the most valuable things about plants is their smell. Not all smell good, of course, but those that do have been used to scent clothes, rooms, hair and skin for thousands of years. Many perfumes come from the flowers of plants such as rose, jasmine and orange.

This Kayapo child from the rain forests of Brazil, in South America, is wearing face paint made from plants.

Plants that cure us

For most of our history, plants have cured many illnesses. In countries like Britain, we now use many man-made medicines. In other parts of the world, people still rely on medicines made from local plants.

The herbalist Nicholas Culpeper (1616-54)

Different cures

Wych (pronounced 'witch') hazel has antiseptic qualities which make it good for cleaning wounds. It was used by the native North Americans to treat many complaints, from back-ache to ulcers. They used extracts from the bark and leaves to prevent swelling and stop bleeding.

The painkiller called aspirin was first made from the leaves and bark of the white willow. The ancient Greeks and native North Americans used it long ago. It has been used to treat pains like toothache, earache and headache. Modern aspirin is now made artificially with chemicals based on those found in the willow.

Oil extracted from the evening primrose is used as a remedy for arthritis, migraine, asthma, eczema and high blood pressure.

Herbals

Centuries ago, before there were doctors and surgeons, some people used a wide range of plants as cures for illnesses. These people (known as herbalists) had a vast knowledge of the plants they used. They put their knowledge into books called herbals. One of the most famous herbals was written by Nicholas Culpeper and published in 1653.

The herbalists had some ideas that modern scientists find hard to accept. They believed, for example, that certain medicines only worked when the planets were in certain positions.

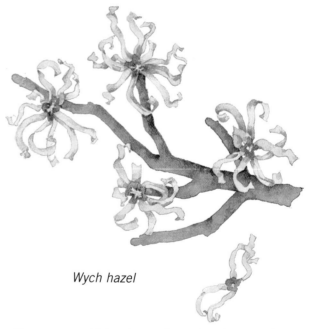

Wych hazel

Every part of henbane is poisonous and yet it has a long tradition of being used as a medicine. The plant contains chemicals used to relieve travel sickness and stomach pain. It has been used to treat some forms of mental illness.

Barefoot medicine

In countries like Britain, we still use some plant remedies, but most of our medicines are artificially made. In other parts of the world, people still use medicines from plants. In China, traditional (or 'barefoot') medicine is part of the country's health-care system. There are several special gardens where around 5000 medicinal plants are grown.

Traditional plant medicines on sale in China.

Evening primrose

Periwinkle was once used to treat inflammations of the skin.

Rain forest medicine chest

The indians of the rain forests have survived for thousands of years using a large number of medicinal plants found in their natural home. One group uses over a hundred different plants. The rain forests are rich in plants that may be useful to modern medicine, but only a few have been examined. Already the forests have given us remedies for malaria and drugs that have helped in the treatment of cancer, diabetes, Hodgkins's disease and snake bites. It is hoped that the rain forests may yet give us cures for other diseases.

Plants we use every day

Throughout our history we have used plants to make our homes, clothes, furniture, tools, games and musical instruments. Artificial materials are now common, but in some parts of the world people still use locally grown plants in their everyday lives.

Plants in the home

Inside our homes are hundreds of things made with plants. There are baskets made of thin willow stems; chairs made of rattan (the stems of a climbing plant from the rain forests); floor and wall tiles made from the bark of the cork oak; carpets made of cotton; paints and varnishes made with plant oils; doormats made of coconut fibres.

Plants for writing on

Another everyday material that comes from plants is paper. Today, most paper is made from wood which has had its bark removed. The wood is made into a pulp by machines or with chemicals. Bank notes, which need to be strong and long-lasting, are made from hemp, flax and cotton fibres.

Plant clothes

Clothes can be made from cotton plants. The cotton fibres are the fluffy white hairs on the seeds of the plant. The process of making fabrics from cotton has been carried out in India since 3000 BC. Linen is a material made from the stems of the flax plant.

Plants that shelter us

Wood is the plant material most commonly used to build homes. It is strong and flexible and lasts a long time. Where there is a good supply of timber, people make whole houses out of wood. In other places it is used for parts such as door frames, staircases and roof supports. Reeds, palms and bamboo are also used to make houses.

The Marsh Arabs of Iraq have built their houses out of reeds for centuries. In the foreground are reed mats ready to be sold.

Plants at play

When we play sports and games, or paint pictures, we may well be using plants. Sports equipment is often made with artificial materials, but cricket bats are still made with wood from the willow tree. The best baseball bats are made of ash wood. Oil paints contain oils of linseed, poppyseed or sunflower seed. Musical instruments are made from a vast variety of woods; some wind instruments use reeds in their mouthpieces.

Bamboo everywhere

Every day, nearly half the world's population uses something made out of bamboo. It is used to make window frames, mats and screens, furniture, fans, paper, water pipes, storage vessels and musical instruments. It can also be eaten! Young bamboo shoots are used in Chinese cookery.

Key

① cane chair
② wicker basket
③ cotton cushion
④ wooden baseball bat
⑤ paper shade
⑥ paper pad
⑦ wooden frame
⑧ cotton curtain

Keeping gardens

All over the world, throughout the centuries, people have had gardens. Often they have been used for growing useful plants, but most gardens are kept for pleasure. In Britain, gardening for pleasure first became popular in the sixteenth century.

Trade in plants

The first ornamental gardens kept only local plants. As transport improved, so people began to trade in plants. Gardeners in Europe in the sixteenth century were fascinated by plants from Asia and America. A botanist went on every new trip. Eventually, plant sellers were sending collectors abroad to find new plants.

This orchid from Costa Rica, in Central America, is endangered due to the trade in wild plants.

A harmful trade

Today, as in the past, many plants are taken from one country to be sold in another. Sadly, sometimes too many plants are removed. They become rare in the wild. In Britain, there are laws against picking wild flowers and plants. We should only buy plants grown in garden nurseries. However, there are many reasons why plants become extinct. Some plants have been saved from extinction because they were grown in gardens.

Gardens for all

All over the world, there are public gardens and parks for people to enjoy. Most people like to have plants near them and enjoy looking after them. People can grow plants wherever they are. If they have no garden, they use window boxes or pots on windowsills.

In this garden in Paris, France, someone has made the most of a little space above some windows.

Looking after the soil

To stay healthy, plants need certain minerals from the soil (see page 8). Healthy soil also has creatures such as worms which help plants to grow. Gardeners look after the soil by putting back the nutrients that the plants use up. This can be done by adding artificial fertilizers, but some people prefer to use natural compost or animal manure. Compost is made up of rotting plant material and kitchen waste, such as carrot tops and tea leaves.

Gardener's ideas

Gardeners are always looking for ways of growing more successful plants and for destroying diseases and pests. Many believe that seeds should be planted before the new moon. The growing moon, as it goes from new to full, is thought to help the plants to grow. Gardeners also believe in 'green fingers'. A person with 'green fingers' is said to be particularly good at making things grow.

Compost or manure need to be dug into the soil, at least once a year, to keep it healthy.

Words to remember
Botanist – a person who studies plants where they grow.

Plants and fossils

Fossils are the remains of plants and animals that lived millions of years ago. The plant fossils that we have found show us roughly how plants have developed over time.

Many plants of the past no longer exist. Others live today, but look very different to how they once were. This is because they have adapted over the years to changing conditions. A few have hardly changed at all.

Changing through time

Among the earliest plant fossils are those of blue-green algae. They lived in the Earth's waters about 3000 million years ago. These plants began to make oxygen and it was this gas that later made animal life on Earth possible.

Plants began to live on land about 400 million years ago. Gradually, they developed stiffer, woody stems which enabled them to reach towards the light. These early plants were quite small, forming forests only one metre high.

Larger plants covered the land about 345–280 million years ago. Ferns, horsetails and lycopods grew to the size of very large trees. Then came the conifers. Flowering plants appeared 140 million years ago. By 65 million years ago the world's forests looked much the same as they do now.

Extinction

Flowering plants are now the most successful group of plants. Many plants have become extinct. Some members of the earlier groups of plants still survive, but they often look different. For example, we still have ferns, horsetails and lycopods today, but they are much smaller plants than the great trees they once were.

Living fossils

Some of these surviving plants have not changed very much over time. They are called living fossils. Ginkgos must have been eaten by plant-eating dinosaurs and they look much the same today.

Many of the modern plants we have today developed between 136 and 65 million years ago. Fossils of these plants can tell us a lot about how the climate has changed since then. For example, we may find a fossil of a plant that today likes damp, hot conditions. The fossil may be found in a desert area. This suggests that, a long time ago, the desert was the site of a tropical rain forest.

Ginkgos are only found in the wild in one tiny area of China, but they are grown in parks and gardens all over the world.

About 300 million years ago, large parts of the Earth were covered with forests of large ferns, lycopods and horsetails. Their fossils form the black 'rock' which we burn: coal.

This fossil ginkgo leaf shows that the plant has changed little over time.

Saving our plants

We owe our lives to plants; we could not survive without them. Yet we don't always treat them very well. Thousands of plants are in danger of extinction. This is usually because their habitats are being harmed or destroyed. It makes more sense to look after plants.

The Earth needs plants

Plants make their own food. They are the most successful and largest group of living things that can do this. Plants are the first link in most food chains; without them there would be no food. Plants also make oxygen in the air. We need this gas to breath.

Plants' roots hold the soil together and stop it being washed away by torrential rain. The leaves that fall from the plants enrich the soil.

Plants are often the first living things to move on to lands that have been damaged. This damage can happen naturally, by the eruption of a volcano, for example, or by the actions of human beings.

Losing plants

Every year, more natural habitats are removed. They are cleared to make room for houses, roads, factories, fields and shops. We also spoil what it left by polluting the soil, the water and the air. As a result, we lose plants and animals.

In some parts of the world, people are trying to stop the destruction and mend the damage. They are cutting down on pollution, putting plants back into their proper environment or simply leaving areas alone so that plants can grow back on their own. They are also trying to preserve what has not been harmed in reserves and national parks. Worldwide, there are many thousands of natural habitats that are now protected by law.

What can we do?

We can all do something to save plants. For example, we can try to be less wasteful by buying less and recycling more. In particular, we can try to waste less paper. We can try not to buy products made from plants that are rare. We can act locally to protect patches of wild land from harm, and we can grow our own plants. The smallest garden can give many plants and animals the chance to live. If you do not have a garden, use a pot on a windowsill (see page 43). Growing just one seed, and tending to the plant that grows from it, is a valuable thing to do.

All over the world, people are planting trees in an effort to give our planet a green future. ▼

◀ *Plants are lost when the habitat in which they live is destroyed. They are also killed by pollution. In this river in Tasmania, the plants have been killed by copper pollution.*

Published by BBC Educational Publishing, a division of
BBC Education, BBC White City, 201 Wood Lane, London W12 7TS
First published 1995
© Andrew Charman/BBC Education 1995
The moral right of the author has been asserted.

Paperback ISBN: 0 563 35538 7
Hardback ISBN: 0 563 39785 3

Colour reproduction by Daylight, Singapore
Printed and bound by BPC Paulton Books Ltd

Photo credits: A-Z Botanical Collection Ltd **pp. 2 (bottom), 9 (bottom), 25;**
Ardea London **pp. 18, 32 (left);** Bruce Colman Ltd **pp. 12, 23, 45 (top);**
Ecoscene **p. 47;** Mary Evans Picture Library **p. 38;** Luke Finn **pp. 17,
36, 37 (top);** Garden Picture Library **pp. 42, 43 (top);** Garden/Wildlife
Matters Photographic Library **pp. 20, 43 (bottom);** Robert Harding
Picture Library **pp. 8, 41;** Hutchison Library **p. 39;** Natural History
Museum **p. 43 (bottom);** NHPA **pp. 3, 9 (top), 16 (top), 29 (bottom), 31,
32 (right), 33, 35, 46;** Planet Earth Pictures **pp. 2 (top), 11, 14, 16
(bottom), 24, 27, 29 (top), 37 (bottom).**

Illustrations: © Martine Collings, 1995 (pages 6, 7, 12, 13, 19, 26,
30–31); © Charles Raymond, 1995 (pages 4–5, 10, 15, 21, 24, 25,
28, 33, 34–35, 38, 39, 40–41, 44–45); © John Shipperbottom, 1995
(page 22).

Front cover: Tony Stone Images **(main picture)** *a fern*, **(bottom right)**
orange gerbera blossom; Bruce Colman Ltd **(top right)** *bumble bee.*